Soul Talk With Cells

SOUL TALK WITH CELLS
WHAT WE REALLY WANT IS TO PLAY

by

Flora Sue Gardner

Strategic Book Publishing
New York, NewYork

Copyright © 2009
All rights reserved – by Flora Sue Gardner

No part of this book may be reproduced or transmitted in any form or by any means, graphic, electronic, or mechanical, including photocopying, recording, taping, or by any information storage retrieval system, without the permission, in writing, from the publisher.

Strategic Books Publishing
An imprint of AEG Publishing Group
845 Third Avenue, 6th Floor - 6016
New York, NY 10022
www.StrategicBookPublishing.com

ISBN: 978-1-60693-953-6   1-60693-953-X

Printed in the United States of America

Book Design: SP

Dedication

To the vibrant health of our magnificent cells

Acknowledgements

Completion of this book would not have happened without the inspiring thoughts from friends, innumerable spiritual and scientific books, and guidance from editors at Strategic Book Publishing.

It is with great pleasure that I acknowledge and appreciate the wisdom and insights over the years as my thoughts came together for this book in conversations with neighbors, teachers, classmates, clients, family members, co-workers, and professional colleagues.

Enormous gratitude goes to my friends at the Inner Light Center where I am privileged to join with them in reverence for all aspects of life and to be continually surrounded by the presence of unconditional love.

Contents

Introduction 11

Part One

My Experiences with Health

Chapter 1	A Personal Picture of Health............................	21
Chapter 2	Cells Are Alive...	27
Chapter 3	Tales of the Food Industry.................................	33
Chapter 4	Know What You Eat..	39
Chapter 5	Shortfalls and Poor Health.................................	43
Chapter 6	Cells Know How to Function.............................	49
Chapter 7	Allow Cells to Choose.......................................	53
Chapter 8	Live and Love Life...	57
Chapter 9	Play...	65

Part Two

The Journey of Desired Change with Reflections on Pictures

Reflections on Chapter 1:
A Vision of Health and its Celebration.................................73

Reflections on Chapter 2:
Investigating the Source of Health, and How to Influence It......75

Reflections on Chapter 3: .
Reconciling the Environment as Critical to Health....................79

Reflections on Chapter 4:
Discovering Who to Trust in the Industry..................................83

Reflections on Chapter 5:
Reversing the Damaged Food and Health System....................87

Reflections on Chapter 6:
Adding to Understanding by Looking Closer...........................89

Reflections on Chapter 7:
Allowing for the Intelligence of Cells..91

Reflections on Chapter 8:
Dancing in Alignment with Love and Life................................95

Reflections on Chapter 9:
Celebrating Health and its Manifestation..................................97

Appendix..99

Introduction

A shock to my belief system woke me up, and what happened reaffirmed my conviction that personal choice is a freedom that I value now more than ever. With my background in clinical nutrition, I had been working in hospitals counseling patients for twenty-five years. During that time, rarely had nutrition been the key factor in a patient's recovery to full health except with diabetes. Instead, surgery or medications were the fundamental reasons for improvement, along with supportive services, until the patient was able to go home. This usual pattern of events was disrupted when a medically diagnosed patient recovered without medications or surgery using a combination of nutrition and lifestyle. Although as a nutrition counselor I was aware of the value of nutrition, I was startled by its significance in this particular case. The patient had taken the lead, had gotten reputable guidance, and had made his own choice regarding his health. In the years since these events happened, I have often reflected on what occurred and how it forever changed my view of life.

As I look back on my own life, it is easy for me to recall that I have always had the ability to choose freely. However, along the way, as we all know, we find there are choices that get a more favorable response than others from those around us. Conditioning during our early years alters the innate intelligence that a child brings into the world, and I was no different. At an early age, I made the choice to gain acceptance and approval, and this choice kept me out of trouble. Unfortunately, actions such as these are a loss for our planet, especially when we all choose to do the same thing.

As an adult, I have noticed that two choices quickly show personal values and say a lot about a person. These are the choices between money and health. Naturally we all want both, as each contributes greatly to our lives, however, with a closer look, each choice reveals evidence of specific values.

Health can be observed fully expressed and displayed in our daily lives in a number of ways, including physically, mentally, emotionally, and socially; and all of these characteristics are developed internally by the individuals themselves. The development of these characteristics can be guided by others, but the work of internal growth is due to the initiative put forth by the individual.

On the other hand, money is an external exchange of energy in the form of tangible symbols; and these symbols have been given a value in order to provide a way for us to live and relate together economically in the world. The values for these symbols are identified by numbers that declare their relative exchange energy; and this structure for exchange is further expanded by number systems declaring changes in the value of goods and services over time. Although this is very fundamentally described, all of us are engaged in this external exercise of money exchange for our livelihoods and well being; and we consistently have the opportunity to use this system with increasing mastery as the years go by during our lives.

Both internal and external aspects of life are critically involved when one ponders the question, "what matters most to me, my money or my full capable function as a human being?" Whereas a lapse in health may not severely hinder the ability to make a living and build up prosperity, the adaptations, and limitations associated with it can affect the options available for making that income; and this makes a difference. It demonstrates that health is the starting point for a person's successful participation in the wide variety of working options in the world; and although education is an obvious variable, learning is generally facilitated by the presence of good health.

In order to develop the unique talents that residents possess in our communities, health must be a primary goal; and every useful talent in the community contributes to the living vitality of the whole. In fact, when we look at the nature of all living organisms surrounding us in the environment, it is the health of the environment itself that supports our very survival.

The priority of health between these two options does not

diminish the value and importance of money. Instead, it enhances the respect that is due to it for the energy it contributes; and it focuses attention on the part it plays in supporting fully functioning individuals, families, communities, businesses, and other organizations that lead us to masterful stewardship in living well with nature and money. Meaningful activity that engages our hearts in our work and in building mutually beneficial relations among ourselves and with nature is work worthy of our time and money. Ultimately, such mastery ensures our survival on this planet and our joy in living.

Human beings have God-given capabilities that allow us to create new ways of thinking and living; and both our level of health consciousness and our expression of individual spirit are mutually beneficial human capacities that urge us to raise the bar for ourselves while exercising our innate ability to rise to the occasion. Applying our abilities outward in the environment includes driving the external energy of money exchange to full creativity.

It is a choice that we are free to make; and a merge in this manner of external pursuits with internally developed human capacities facilitates a union and balance of energies that opens the door for creating at higher levels. Indeed, our natural tendencies toward technological creations and highly structured systems may yet manifest a system that congruently supports, sustains, and renews health and all the gifts of Nature at the same time.

Already, many of our manmade inventions have helped us communicate and understand the diverse and rich cultures of other human beings simply by having more information readily accessible electronically; and, in a general sense, this system includes our system of money exchange. First, we created coins, notes, and checks; and this was followed by cards and account numbers; and that facilitated electronic transfers to move money in seconds between buyer, seller, financial institutions, divisions of government, and all types of business organizations. These advances for transferring money changed our thinking about money; and it also changed the way we related to money, its energy, value, and wisdom in using it.

As manmade inventions for transferring money energy continue to be introduced, we have a responsibility to assure congruency with the natural order of life in these critically important systems.

With health for the environment and everything that lives in it as a desirable goal, then feasibility becomes an issue; and this is particularly true with regard to our current practices and beliefs. Over the years, as a practical matter, I have observed that individual lifestyle greatly influences the choices made between money and health. Issues of genetics, culture, gender, age, history, education, and a host of other variables make individual choices distinctive. Those who learn at an early age to take good care of their health and value the practices that contribute to it have an edge, as a healthy body supports fuller participation in the opportunities of life.

While it is true that we are living longer, statistics show that increasing numbers of people are experiencing long-term physical limitations. These long-term problems are lifestyle diseases, meaning they are manmade and virtually unheard of in the early 1900s. Our modern age has changed the way we live, and diabetes, heart disease, stroke, cancer, osteoporosis, arthritis, and Alzheimer's disease are all examples of long-term lifestyle diseases.

As a Registered Dietitian with a Masters Degree in Nutrition, my training included therapeutic nutrition for patients in clinics and medical centers. I spent twenty-five years counseling patients and families about modifications they could make to their diets to help relieve the symptoms of limiting health conditions. For years, it was work that I loved, and many patients were pleased with both the changes they chose to make and the results they experienced.

However, as times changed, I began to see another phase of nutrition that was more exciting. Although therapeutic work in hospitals is critical, and there will always be a need for medical specialists, it became obvious to me that many of today's diseases are preventable.

In the early 1990s, my interests changed, and I began to devote my attention to the prevention of health problems before symptoms occur.

Response to disease is much different than a focus purely on health. They may sound the same, but there is, in fact, a distinction. The former calls for reacting to situations that have already occurred, the latter has a proactive aim to promote a desired effect. Self-reliance, self-trust, and resourcefulness through accessing credible information are synonymous with prevention. In other words, prevention requires individual responsibility for, and confidence in, personal choices. At the same time, it includes maintaining an awareness and openness to the advancing knowledge of the highly trained medical professionals who are dedicated to caring for the body in diseased states.

My training was in disease care and I had the privilege of administering state-of-the-art nutritional guidance in clinical settings. During the latter part of those years, I was introduced to a man in a social setting who was a professional athlete and appeared to be in good health. Living with the discipline and physical extremes required by an athlete's lifestyle, taking good care of his body was typical and he continued the practice even after he chose to enter the business world. However, in the business world where he worked, the stresses of stringent productivity and constant travel made his health objectives challenging, as his schedule was often unpredictable and hours were long.

Several months after our introduction, my friend told me that, after experiencing some sharp pain in his abdomen and seeing a doctor, he had received a diagnosis of pancreatic cancer. Having worked in hospitals for so many years; I immediately knew what that meant. Pancreatic cancer was a disease that always made me cringe, having seen many patients die from it. It was devastating to know that my friend could expect to live for six months or less.

Surprisingly, my friend was not devastated, as he had an outcome other than death in his mind. After rejecting his scheduled invasive surgery, he went about conferring with pancreatic cancer specialists and progressive research doctors and saw the possibility of recovering using modalities such as innovative nutrition, meditation, stress management, and total body awareness. He became part of a group of fifty participating pancreatic cancer patients for a medically recommended and administered project conducted through a major medical center in the United States. My friend was delighted, and I was amazed. He held on to his athlete's perspective and, at that point,

I was a silent observer.

For approximately two months, I watched as my friend took seventy-two pills every day. Yes, we counted them. Two-thirds of the pills were nutritional supplements and the rest were pancreatic enzymes. None of the pills were pharmaceutical medications. He received no chemotherapy or radiation. It was living on the edge as far as I was concerned, but he was disciplined, compliant, and focused. He integrated meditation, stress management, and total body awareness with a Mediterranean-style diet and the nutritional supplementation. After several months, when they tested him repeatedly for the presence of cancer, they found none.

It was after my friend's test results that I knew my direction in nutrition was forever changed. From then on, I decided to study more about preventive nutrition and the innate vitality of the human body.

In years past, my therapeutic nutrition training had included biochemistry, nutrition, and physiology, along with classes in microbiology where I closely studied cells through a microscope. While much of my newfound confidence in prevention continued to be based on research information from these fields, I knew I needed to add the study of immunology to my training. Without gaining an appreciable knowledge in this field, it mystified me as to how vitamins and minerals, at such unexpectedly high levels, could have such a startling effect on health. From my friend's experience however, it indeed appeared, and unmistakably so, that cells could recover to a natural state of health when they received adequate support to do so.

I was already familiar with the details of nutrition such as calories, proteins, fats, carbohydrates, vitamins, minerals, antioxidants, phytochemicals, extracts, and other subnutrients, I was also experienced with individual food habits and preferences. What I lacked was firsthand experience with what I considered unusually high levels of nutritional support for cells.

For my own research, I decided to pursue this experience in my own life. I knew my diet was not as good as it should be, especially since, as a healthcare professional, I "knew better." My calorie intake was lower than what my body needed, and because of this, I also wasn't getting the nutrition that I needed. Honestly, I kept so busy that I had not stopped to think about it, and my lifestyle seemed too unpredictable and spontaneous to plan and prepare balanced organic meals every day.

I made the decision to try the same supplementation program used by my friend, formulated from scientific tests during which the cells demonstrated the exact nutrients, and amounts of those nutrients, that supported their health continuously. I refer to these demonstrations as "cell talk."

Generally healthy, I chose to use normal amounts of the same supplements, reasoning that if diseased cells became healthy with this specific combination of nutrients at high levels, a more normal dose was indicative of increased good health for me. I used the supplements as directed on their containers and did experience improvements in my health. In fact, I noticed such a difference that, for me, it was proof that cells do thrive on full nutritional support, and from this, naturally, the whole body benefits.

Although I had evidence, at this point I was still without understanding. It was apparent that cells respond well when attention to their needs and functions was paid, this seemed obvious and respectful, but at the same time I was still mystified and curious.

This mystery was the inspiration for this book. Rather than remaining committed to the repair and recovery aspect of attention to the body, and the disease approach to care of the body which is based in treatment and rehabilitation, I chose to work with constructive ideals that can lead human beings to a point of harmonious living with vibrant organisms, most especially our own microscopic cells.

Soul Talk With Cells: What We Really Want is to Play offers this different perspective. It is an idea that honors aligning with the natural potential of living organisms to create order and perfection. In fact, our thoughts, attitudes, and awareness of our body and mind affect our every move. These thoughts can be inspired, positive, or negative and certainly, in any case, these thoughts help lead to outcomes. We have the choice of aligning our thoughts and actions with the natural, and healthful, design of our bodies and our planet.

In the following pages, I describe what I have learned. This is not an in-depth discussion on the subject of nutrition and the body, and it is purposely not a technical presentation, as there are many well-written books that cover that aspect of nutrition and the body. My purpose here is to present the perspective of cooperating with our body's cells, and I believe that how we think about life does make a difference—in some ways, it makes all the difference.

My hope is that readers will find this viewpoint valuable and helpful. During my study, I learned to accept cells as they actually are and to trust their God-given nature and intelligence. This meant relinquishing an egotistical tendency to overtake their functioning with a "better idea." After my investigation, choosing to deliberately align and cooperate with cells became an easy choice.

This book is presented in two parts. Part One describes my experience with nutritional practices and personal attitudes that eventually manifested joyful health and play. Part Two is the same journey told with pictures and my own reflections related back to the chapters in Part One. I offer Part Two as an opportunity for readers to ponder the images. The soul, and body, language within the scenes may communicate more deeply when individuals apply their own insights, imaginations, or meditations. Having selected the pictures, I offer only possible reflections, and readers may find valuable interpretations based on their own personal experience.

PART ONE

Part One is an expression of my journey,
filled with lessons and discoveries,
from a background in clinical nutrition to
joyful health and play.

Chapter One
A Personal Picture of Health

Each one of us has mental pictures throughout the day that make a difference in our lives. For example, if the gas gauge in our car happens to show close to empty, we see or envision ourselves going to the gas station for more gas, and we make that vision real with action. The thought occurs before the action and then, naturally, leads to the action. If we are playing a game, we envision the next move we want to make in order to win the game, and we then make the move accordingly. Similarly, if we are hungry or thirsty, we envision getting something to eat or drink. Examples from everyday life are endless. Before we take *any* action, we have a thought or picture in our mind not only of the action, but also of the positive result of the action. Naturally, it is a problem if we unwittingly picture a negative outcome, possibly thinking that it doesn't matter what we picture. It does matter, however, because this is how it works—a thought leads to a manifestation. So, indeed, the thought does matter, every time. If we picture a negative result, we are more than bound to get it.

As children, we are subject to the visions of parents, teachers, friends, and other authority figures as we have yet to reach the point of understanding and assuming responsibility for our own thoughts. However, as we grow and mature, we have the opportunity to recognize that we have choice. After a number of years of living on this planet, evidence begins to show that our visions and thoughts do make a difference to outcomes. Like it or not, our individual thoughts often influence the outcomes that we complain about. It should come as no surprise then that good health can only happen when it is pictured in the mind.

If you have any doubt about this, take a moment and consider your own responses to the following. Think of an activity in which you recently participated. For example, the activity of picking up this book to read it, or the activity of walking into the room to locate it, or simply the activity of getting out of bed this morning. Any of these, and the many others that come to mind, is fine.

Now, take a moment to ponder the image in your mind that preceded picking up this book. Was it an image of picking up the book, finding the book, or something similar? Perhaps, before you walked into the room where the book is located, you envisioned walking into that same room. Perhaps, just before you got out of bed, you had the thought, "I'm going to get out of bed now."

This concept is important to you because if you want good health, you must learn to picture it. All of your follow-up actions directly relate to your mental picture—so make sure to have a good picture.

It is an understatement to say that both physical bodies and thoughts of the mind, sometimes called body-minds, are complex and amazing. Could there be a more worthy pursuit than operating both the body and mind at full function? What we envision for our health primarily benefits ourselves. However, our hopes and dreams are also of interest to family and friends, as well as the enormous and expanding wellness industry.

Broadly described, the wellness industry includes all providers of services and products that support the well-being of the body.

A short list of entities that support participation in the activities of life includes food and food service, of which agriculture and farming are part, medicine and pharmaceuticals, which includes the personnel, alternative medicine, fitness, personal care for both skin and hair, nutritional supplements, health insurance, divisions of state and federal government, and private businesses.

Many livelihoods are engaged in wellness and, because human

lives are at stake, the importance and economics of this industry are huge.

At the heart of promoting optimal health is gratitude for life. Until we look closely at the gift that life is, we have not really noticed what is actually happening. It is when we become aware of life at the most fundamental level that we can develop a genuine appreciation. With human beings, our cells are the basic level of functioning, and an awareness of cell activity is of primary importance to responding appropriately to the needs of the entire body.

Although science is not everyone's favorite subject, evidence provided by scientific investigation gives the clearest picture of topics and issues that are studied. Repeatedly, these results are so clear that we can rely on them every time and, in fact, when the studies have been well performed, they reveal the truth of any matter at hand. Such details help to explain our world and make public facts that would not otherwise be available. Because of scientific investigations, our understanding increases.

Scientific work with cells, as they are not visible to the naked eye, is done with the aid of microscopes. As we study cells in this way, we learn their functions and gain an understanding of their importance and relationship to everyday living. In return for their valuable function, humans can create a relationship with our cells that is supportive of their needs and, in doing so, we can take a quantum leap forward towards better health.

How do we go about taking this quantum leap if that is what we choose to do? The essential first step is a vision of good health. A close second is education, and as our purpose here is a focus on vision, only this brief note on information follows. Scientific information regarding all types of cells (human, animal, and plant) is readily available in libraries, through the media, and on the Internet, and if we want details, we can find them. As a word of caution, some scientific reports are more exhaustively researched than others are and, unfortunately, some sources are just plain misleading for a variety of reasons. The purity of data is critical and completely dependent on the research methods.

Listed below are three verifiable fundamental descriptions of cells.

1. Every cell of the body is a living unit with energetic and active internal parts.
2. Each cell functions better when there is a response to its needs for life.
3. A mutually supportive relationship between humans and their cells benefits the whole body.

These are the basics, and additional information provided in later chapters will align with these facts.

Because we are all unique, everyone has their own vision of the world and of themselves within it. A vision so strongly desired that feelings and emotions arise related to living that vision in each moment is powerfully driven. When we know what drives our vision, it promotes enthusiasm for it. Examples of motivating reasons for good health are everyday desires such as participating in family gatherings, successful performance on the job, self-reliant mobility and care, sporting events, athletic competition, and many others. Since a vision requires action to make it real, we need to understand what is important enough to us individually to move us into action. Insufficient drive related to good health is costly in both the physical limitations and the financial payout that poor health brings out.

It is beneficial to take a moment and write down the reasons you desire good health. It may be that you see yourself enjoying sculpting, painting, playing a musical instrument, traveling, running for political office, gardening, taking care of children, or starting another career.

Whatever you want to experience, write it down on a sheet of paper or on the lines below.

1. My vision for my body is

_____
_____
_____

2. Motivating reasons, desires, feelings, and emotions driving this vision are

_____
_____
_____

3. Happenings and activities that fulfill this vision are

_____
_____
_____

Now that you have a clearer picture of what your own good health looks and feels like, some options to manifest this vision are in the chapters that follow.

However, first we must review some very basic science.

Chapter 2
Cells Are Alive

    Even though cells are vital to our lives, we cannot see human cells without a microscope. The largest human cell, the ovulated egg in females, is approximately the diameter of a single human hair, and all other cells are smaller. The easiest cells to see as a group and touch with our hands are those in the form of skin; and skin is the largest organ of the body. Microscopic cells have first formed together into a specific type of tissue; and then similar types of tissue have formed into the organ, which in this case is called the skin. This is the same way all organs are formed like the heart, lungs, bones, muscles, and others.

    There are twenty major kinds of human cells and many subcategories and all these cells have specific functions. Depending on the size of a body, one hundred trillion cells are functioning twenty-four hours a day, and every one of these cells needs good nutrition.

    Unfortunately, humans often take the function of their cells for granted. Somehow, in our busy lives, we have developed the mistaken idea that cells do not need our awareness. While it is true that they function on their own to a certain extent, they do need our cooperation to function optimally. Their cooperation is on our side by design. When we work together with their needs, rather than separate from them, our interactions with them are healthful. A thoughtless attitude has the opposite effect—one of neglect. Eventually the signs of neglect will show up in the body and, generally, limited functioning is the outcome.

Humans regularly challenge the cells of the body to function under extreme demands, and, amazingly, when they aren't totally damaged in the process, they adapt. When we treat the same cells more favorably, they come back to full functioning.

Expecting cells to function without our cooperation the majority of the time is similar to expecting mechanical devices to serve us well continually without any maintenance. Three examples of this concept are televisions, cell phones, and automobiles. How they work is not as important to many of us as long as they do work.

Let us take a closer look.

Television: The operative parts of a television are understood by specialists who have been trained to assemble, repair, and improve television technology. That is lucky for us because when we turn on the television, we are not paying attention to what makes the picture or sound come in perfectly. Instead, we are interested in getting the program we want tuned in from the remote. The rest just happens; and we depend on it happening so we can keep informed and entertained.

If for some reason a problem arises, we call the repairperson who understands the technicalities. We have only three responsibilities to cooperate with the television and promote proper functioning; checking the batteries in the remote, checking the power cord for connection to the wall socket, and making sure the power bill is paid. The rest we take for granted.

Cell Phones: Although these handheld devices are much smaller than a television, they usually have many more functions. They are not just telephones, but computers, calculators, cameras, video and voice recorders, and some have several other record keeping operations as part of their communication system. We have come to depend on this high tech equipment, but we are generally too busy to think about how they work every time we use them. We focus on using them to accomplish an intended purpose, only acknowledging the technical aspects when we are learning how to use a new phone. After mastering a few functions, we relate to them dependably.

The only cooperative responsibility we have, other than paying the monthly bill, is charging the battery every night. As amazing as cell phones are, they still require power. If a phone's functioning stops, charging the battery, paying the phone bill, or calling the technician solves the problem. The rest of the advanced operating system that causes the phone to work we generally take for granted.

Automobiles: Owning and maintaining a car is a responsibility nearly everyone accepts once they have the ability to make the purchase. Awareness of the operation of an automobile engine and all the working parts is more familiar to most than the parts of a television or cell phone. We find out quickly what happens to the full and safe operation of an automobile if we are negligent with the brakes, the headlights, the oil, the gas, the windshield wipers, or any other critical component of its system. Once we find ourselves stranded or lose control of a vehicle, we realize we have some responsibility, and we are endangering the lives of both ourselves and others on the road.

As with our bodies, we can allow an automobile to function at less than full efficiency. For example, with a car, if we allow the air filter to clog with dust and dirt, the engine still runs and related parts of the engine adapt to decreased air supply. We may not even notice that the car burns more gas because there is less air getting to the engine. The fact that we don't notice this adjustment relates to what happens when we don't take good care of our bodies. Our bodies adjust in this same way, functioning less efficiently.

In fact, we may be more aware of a fault in the functioning of an automobile. We don't like getting stranded unexpectedly and we notice this immediately as a decidedly unwanted problem. Even the possibility of getting stranded keeps us in a maintenance mindset, far ahead of problem development. Sadly, our bodies may not get this same vigilance. Somehow it is easier, or possibly more obvious, to recognize that the small and essential parts of a car are vital to dependability and operate at their best when they get attention from us.

In our busy lives, travel from place to place is usually required and it goes more smoothly when our vehicle, both automobile and body, is in working order. As parts work together for a common purpose, they are demonstrating a form of interdependency and cooperation.

Bodies: Certainly, humans are engaged in an interdependent relationship with the cells of their bodies. Healthy cells can give us their best performance, and cells need our attention so they can provide their best. Cells are with us twenty-four hours a day, seven days a week and, not surprisingly, they are the very form of life in which we live.

Since ancient times, it has been known that a life form serves as the lodging for the soul. More particularly, life, or the essence of who we are, is our spirit rather than our form. This is a distinction well received and acknowledged by many today. Human bodies are the most highly developed and gifted form of life on the planet, and it is fundamental to our nature that a being can witness, and be aware of, the body form and see it as made of indescribably phenomenal cells. This natural ability allows soul, or being energy, to join naturally in *oneness* (no separation) with the essence of our cells. Even though at times we are aware that the function of these cells is constantly serving us throughout the day, we often get busy and neglect their needs.

Interdependence occurs not only between the cells, but also between the organs and tissues. Networking and communication are functions throughout all aspects of the body and to have a fully functioning matrix of communication among our cells, we need to be a conscious part of that network. Eventually, how well we cooperate with our body shows, and sometimes it becomes obvious immediately.

Cells in a healthy state harmonize with life in its natural form. By nature, they cooperate when given the opportunity. It is we, as human beings, who are inclined toward struggle and strife. Indeed, if harmony with life *is* life, then many of us are in discord a good part of the time and missing the full experience that is available to us. Such discord is a reminder for us to get back into life *as it is*. A nourishing environment within the body allows cells to duplicate themselves naturally in fully developed and competent form. We can support that with cooperation by providing the nourishing environment in which they thrive. Indeed, where else would this environment come from? It is our interdependent responsibility, and without our participation they make do with what they have. The action we take to provide for their needs returns a response from the cells based on what we deliver.

Let us take a look at cell activity.

First, cells have different shapes and appearances according to their purpose, and yet, when we look inside different types of cells, we find they all have similar active components. Each cell and its internal elements require nutrition twenty-four hours a day to function optimally. Thousands of interactions occur during every second of life inside a single cell. Each interaction facilitated, in large part, by the energy of nutrition. Although it defies description to enumerate every interaction inside a single cell (try describing how the universe works), observations show that when optimally nourished, they thrive. Indeed, when the cellular environment in which they live supplies the nourishment that enables full health, cells replace themselves in complete health; and, naturally, if necessary components for that process are missing, they adapt and replace themselves accordingly, in which case the outcome is not as complete.

When cells thrive, they support the health of the whole body. The following are three very basic activities of the cells:

1. They create their own energy from what we give them in nutrition.
2. They interact with other cells of the body.
3. They replace themselves consistent with the health of the environment in which they live.

No matter how a cell's activity is described, it doesn't even scratch the surface of identifying all that actually happens. Our disease care system offers many examination techniques for diagnosing the status of our health, and although these techniques are continually advancing and reveal very useful information, they are not able to detect health at the cellular level for *all* cells. It therefore remains that individually initiated awareness and support of our health on a daily basis is a more responsible way to promote health.

With regard to promotion of cell health, there are two questions to answer, what is optimal nutrition as determined by the cells, and how do we cooperate in order to supply this nutrition?

CHAPTER 3
TALES OF THE FOOD INDUSTRY

    The food industry in the United States and worldwide is an important subject, described and analyzed by professionals working in the field and by authoritative commentators in the media. It is a constantly changing, multibillion-dollar industry that affects our health every day. Major influences from new scientific evidence translate into marketing and product content changes, and health claims responding to popular demand additionally influence content and advertising.

    In fact, when we look at consumer demand, our buying habits have been a driving force behind many modifications in the industry. For example, new farming methods now supply produce that is bigger, greener, and more abundant. Initially, this change was thought to be a good one because it not only produced crops faster for buyers, it also resulted in more money for producers. However, in the end, the chemical treatment of the soil that caused the rapid growth of the crops decreased the nutritional quality of the soil. Further, due to the decrease in soil quality, the nutritional quality of the crops grown in the soil also decreased. With more crops being produced faster, methods of harvesting, storing, transporting, processing, packaging, labeling, and other activities to have food products more available also changed. Economically it was beneficial. As long as consumers were buying, producers continued to produce.

The issue of choosing money over health or health over money enters here, and consumers and producers are both involved in this priority decision. Consumers can only spend their money on food products if they are available to purchase, and producers can only afford to make products available if consumers will buy them. The determining point for this important issue is one of education and understanding regarding support for the health of cells, and the advertising of products does not qualify as education. If we remain uneducated and negligent on this issue, we head in the same direction as civilizations from the past that collapsed from ignorance. History has shown that choosing separate ambitions over the health of the environment and those who depend on it leads to the destruction of plants, animals, humans, and finally civilizations themselves. Choosing health first is a choice for cooperation with the cells of our bodies and, ultimately, the cells throughout the environment as a whole. Indeed, the choice for health is a step toward relating with ourselves, and the vibrantly alive cells of our bodies, in a functional way.

Obviously, food products from the farm to our digestive system, and every process in between, affect the health of our cells. Following are descriptions of production activities that have changed the nutritional quality of our food supply. Awareness of these alterations can benefit the reader in choosing foods that best support the work of the cells.

Food from the Earth

All food initially comes from the earth, whether from land or sea. Small fish eat water vegetation, big fish eat little fish, and so on. On land, animals eat the seeds, nuts, and leaves of the land or smaller animals, and omnivores eat all of these. The condition of our soil and water is the starting point for the quality of all our food.

An unfortunate fact is that humans are just beginning to appreciate this. We have not attended to the value of our soil until recently through organic food production. With shortsighted good intentions, we altered our agricultural viewpoint about fifty years ago and added chemicals to our farming methods. The need to maintain the natural nutrition of the soil in order to maintain it in the plants was not given due consideration.

The first time this situation came to my attention was in 1994. Concerned about the nutritional information I gave to the patients I counseled, I interviewed a nutrition professor at a well-known agricultural university to find out some details. I wanted to know what research was underway that would substantively expose this developing problem. The professor told me, "No grant funds are being offered to study the nutritional value of crops, the money is all going to identify methods for growing crops bigger, faster, and greener."

Once again, the priority was economic. Crops were being grown to be more appealing and abundant at market, but what was being sold was of less value to health. Human bodies and cells were denied their needs as their importance took a backseat to commerce.

This is a problem created by humans. The good side is that since the issue is man made, we can correct it, and the first step to this correction is to become aware of what is involved.

Composting, the returning of leaves, grass, vines, and manure to the earth, has historically been the way to maintain nutrients in the soil. Years ago, cropland was allowed to rest, or lay fallow, between years of production as part of the natural soil enrichment process. In addition, the practice of planting different crops in the same field on alternate years also helped to keep the soil nutrition balanced.

Although organic farming today has begun to return richness to the soil and nutrition to the crops, it has also elevated food prices. As a result, many cost-conscious or cost-limited consumers continue to purchase foods that appeal in appearance and with lower cost. These consumers may be saving money on the front end, only to spend more on the back end through issues connected with poor health.

Soil experts say it takes many years, possibly hundreds, to bring soil back to its original richness. In the meantime, the nutritional needs of the cells of plants, the animals eating the plants, and the humans eating both are becoming more and more depleted. Actions to replace the lost nutrition until the soil can replenish itself are critical.

Processed Foods

Twenty-five years after the introduction of chemical fertilizers in 1948, processed foods began flooding the market. These processed foods were highly refined, preserved, frozen, extracted, premixed, ready-to-use, or otherwise altered from their original state, and included such otherwise natural items as fruits, vegetables, grains, beans, seeds, nuts, meats, dairy products, and beverages. Initially, these innovative products were regarded as imitation food items. However, as the word 'processed' became more acceptable, whole lines of these food products appeared.

Today, additives listed on the labels of some processed food products actually alter the taste and quality of a food enough that it becomes difficult to detect what the original taste might have been. Sadly, most of the foods found in supermarkets today are processed, and we have lifestyles that call for this ready-to-eat fare. Unfortunately, although these convenience items are in high demand, any of the most popular food and beverage brands are detrimental to wellness.

In supermarkets, it is easy to find long aisles of similar food items with marketing claims declaring that their product supports good health. Such processed, functional foods can claim support of health based on research showing the advantages of consuming a food component of the product. Choices can be daunting due to the large number of products. In fact, in some cases, if we aren't prepared by knowing what we want before we shop, it can be downright overwhelming. Buying the cheapest products is sometimes the easiest strategy because it gets shopping done quickly, however, by simply making the choice to purchase fewer processed foods, and fresher, more nutritionally valuable foods, we can also help relieve any shopping frustration.

In support of processed foods, they can be valuable when used intermittently as they add variety to weekly menus and interesting flavor adventures. However, 'intermittently' is the key word here. Fresh foods, even with the soil being depleted, are still of greater nutritional value because they have not had to withstand the effects of processing. Minimally processed foods, close to their original form, are a better choice than ready to eat packaged foods.

No one is forcing our food choices. Marketers urge us to buy, count the statistics of what we buy, and then report the numbers to producers for more of what sells, but when we make a purchase, no one is watching us. If we choose to select food items that don't support our health, that is the choice we make for our body, and our cells perform according to the nutrients they are supplied.

CHAPTER 4
KNOW WHAT YOU EAT

The best news about the food supply is that we make the choice as to what we eat. Thoughtfully selecting the foods we eat leads to enjoying and feeling good about its nutritional support for our bodies and minds while it builds our self-confidence and trust. Meals that refresh us with savory flavors and nutrients for our health enhance our mealtime experiences and our feelings of well being.

Let us take a closer look at food selection.

With the numerous activities and distractions of our busy lives, we often opt for convenience simply because it is easier, and yet, the responsibility for eating what we truly want among all the offerings, is ours alone. It helps to know the value differences between foods, and because food origins and the contents of combination foods change, staying informed by learning about products is essential. Where we choose to be ignorant and avoid the responsibility for our choices, costly alternatives and limitations present themselves. Routinely asking simple, sensible questions can help acquire the habit of informed choosing.

On an individual level, listed below are questions to be aware of and ask about foods during your screening process, before deciding whether or not to purchase them. Responses to these questions are a guidance system toward the nutritional support of healthy cells.

Screening Questions

Following are elementary, and yet very important, questions. Designed to collect facts, as only correct information is useful; you no longer need to assume or guess. The answers must be dependably correct to make good decisions for your health. There are four main questions and six that are related. The process is simply asking questions to sort between food items.

What do I screen and why?
Our cells are affected by all foods and beverages, so screen them all. After a while, you will only need to screen new products on the market. A food that supports the health of cells makes a positive difference. The less processed a product is, the more fresh it is and the more nutritionally valuable. Screening allows you to collect the information that will help you decide what to buy. Once you have the information, the choice is yours.

What questions do I ask about foods?

Fresh foods:
1. How long has it been in storage? Time in storage depletes nutritional value.
2. Was the food harvested before full growth? Nutritional value is greatest at full growth.
3. What was the nutritional condition of the soil in which it was grown? If soil is missing nutrients, then the plants are missing nutrients, and the feed for animals is missing nutrients as well.

Processed foods (any packaged food or beverage):
1. How close to the original fresh condition is this food? Refining, cooking, carbonating, artificially sweetening, milling, and all other types of food alteration and preservation deplete the nutritional value.

What do I need to know about labels? The food industry competes for our money. This is good and makes screening practical. Information on the label is carefully selected to appeal to our purchasing. Think beyond the label and advertising messages and ask:

1. Is this a company I trust with my health? The food industry is too large to be monitored by government officials except by an "alert system" which goes into effect after a problem occurs. Labeling and manufacturing standards are set at the company level. Choose a company you trust to label their products accurately and to stand by their labeling if challenged.

What do I need to know about nutritional supplements? The nutritional supplement industry also competes for our money. While this is good, it also makes screening essential. In the United States, supplements include vitamins, minerals, herbs, subnutrients, plant oils, and other plant extracts. These are all processed products, the original being the plants from which the items were developed. As with the food industry, labeling and advertising are carefully selected to appeal to our purchasing. Think beyond the labeling and advertising messages and ask:

1. Do I trust the health of my cells to this company? The supplement industry has been based in guesswork during most of its years in existence. Many approaches to looking at this are available. Two of these approaches are, 1) that nutritional supplements have been manufactured and consumed in forms and amounts based on reports from those who have used them and the outcomes they experienced, and 2) that supplements have been manufactured and consumed in forms and amounts shown to avoid deficiency symptoms in the body. In neither case are supplements tested directly with live human cells in order to observe a response. As a result, most products in the industry are not backed with the confidence that they are providing the basic needs of cells for their health. However, products without that confidence do not harm the cells except by neglect or by causing an imbalance, and typically, people do not die because of using them. Because there is no immediate harm, the government "alert system," which is in place to detect problems in the industry, does not have any evidence to prevent their sale. It is a situation

where the buyer must be aware. Indeed, very few supplement companies direct their attention to the specific needs of the human cell. However, there are scientists available who focus directly on providing the cells with nutrient formulations shown by the cells themselves to be a combination on which they thrive. Since formulations like this are manmade, it is important to make sure the organization producing the supplement is one you trust to manufacture and label their products using the highest standards. For example, they guarantee that what is on the label is actually in the container. Because quality supplement production is highly technical, manufacturing companies that understand the cell's nutritional need for exactness in the content of the supplement assure you of the active ingredients; in doing so, they guarantee the purity and potency of their product so you can depend on the contents as listed on the label.

These questions, and the reasons for asking them, need to make sense to you personally before you can ask them effectively. Take into consideration how knowing the answers will benefit you and your own health.

Cells deserve your contemplation about how they best support your body when they get what they need. The point of screening is for us all to choose responsibly and cooperate effectively with the function of our highly vibrant cells.

Chapter 5
Shortfalls and Poor Health

When inadequately supplied with what they need for a long enough time, problems arise within our cells and this shows up as poor health. Unless we want such problems, we need to take action.

Of primary importance, is our attitude. Cells deserve acknowledgement for what they do, and who would argue with the value that cells contribute to our lives? Where would we be without them?

Details describing the digestion, absorption, and utilization of nutrients both inside and between cells are important to appreciate, and numerous books and journals in medical libraries readily provide this information. Fundamentally, it is important to understand that just as nutrients of all kinds work interdependently among themselves, cells also work within interactive networks. Cells absorb nutrients and then perform transformations that create health. Even one missing nutritional element can stop a specific action as they all are needed in certain amounts. Each nutrient has a distinctive chemistry and function which makes a difference to the performance of the whole, and this is the way it is by Nature's design.

Although the physical energetic needs from food are obvious, for example, we have physical signs of being and feeling hungry when we have gone long without food, cells have needs from emotional, mental, and spiritual states as well. Since emotional, mental, and spiritual energies also affect our cells, it makes sense to be aware of this and practice attitudes and habits that create well-being. Cooperation in this manner extends an additional energy to cells that supports them with the energy of positive thoughts and behaviors, and when they receive such positive energy, they respond positively. As a fact, negative consequences occur in the absence of cooperation, and we feel that disharmony as struggle, strife, suffering, and poor health. Lately, an increasing number of us have experienced a great deal of disharmony, and this has shown up as diseases in the body.

Increasing Numbers with Degenerative Diseases

It used to be that infections in the body caused our biggest health problems and, fortunately, when medications were administered, our bodies recovered. However, in the last fifty years, that picture has changed as we now have diseases that stay with us throughout life. Heart disease, diabetes, osteoporosis, cancer, stroke, arthritis, obesity, and many other diseases fall into this category. Even when well managed, they require cautious participation in life's activities, and living with disease has become a lifestyle for many.

The hardest piece of news about degenerative diseases is that we brought these problems upon ourselves. Poor choice of priorities, general disharmony, ignoring long-term effects, abuse of the soil or water, or any other approach that is adverse to Nature's design could be the origin. No matter the original cause, our understanding has advanced far beyond previous beliefs during the last ten years. With the new awareness that we are each uniquely involved in our own way, it is obvious that action to reverse health problems is an individual initiative.

Lifestyle modifications are not usually welcomed with open arms. We all like to have free choice over our lifestyles. However, red flags signaling potential poor health mean we should rethink, and possibly rebuild, areas where there is damage or neglect. The ultimate outcome of a lifestyle of neglect is suffering.

I know about this road personally. Without realizing I was choosing it, I was on it for a long time. I suffered for years until I was so tired of discord and health issues in my life that I was willing to do anything to get on a new track. The old saying "an ounce of prevention is worth a pound of cure," turned out to be true. I needed to stop putting temporary fixes and momentary cures on the problem, thereby prolonging the problem, and start identifying the cause so I could initiate a solution. I was the cause of my own problems and that is a tough piece of news to absorb.

Turning myself around began when I started investigating the truth behind my thoughts; books written by author Byron Katie were very helpful with all types of issues. I discovered struggles have many origins, with poor nutrition being only one possibility. Learning to harmonize with the design of Nature the way it actually is, including the design for human beings, works to neutralize potential struggle situations. Living habits that promote health and happiness from all perspectives of life are very much worth investigating and initiating, as they make life a much more fun place to be.

Degenerative diseases that stay with us throughout life usually develop over the course of many years, starting with a few cells and then growing. Recent scientific research has shown that oxidative damage may be an underlying cause. While we need oxygen to live, we also form free radical oxygen—an oxygen molecule that is missing an electron—from pollution in the environment, stress, excessive exercise, and fatty diets. Although our bodies provide some antioxidants—the neutralizers of free radical activity—our environment and lifestyles tend to overtake this natural protection. As a result, we fall short of the neutralizing activity the body needs and excess free radicals steal electrons from healthy cells. These formerly healthy cells then lack an electron and must steal an electron from other cells, and this causes a chain reaction of damage unless antioxidants, which have extra electrons, are present.

Although scientific books fully explain this process in the body, an everyday analogy that is visible to our eyes follows. When cells are altered by oxidative damage from free radicals, it is similar to what is seen on apples when they are cut, exposed to air for a few hours, and they turn brown. That is oxidation. The rust seen on cars is also due to oxidation. Oxygen in the air is chemically altering the original condition and it shows up as browning or rusting. The antioxidant that neutralizes the effect on apples is ascorbic acid, which is found in many citrus fruits, and various commercial treatments neutralize the rusting of automobiles.

Support of cells with good nutrition involves the presence of essential nutrients as well as adequate antioxidants, and the evidence shows that damage occurs without them over extended periods of time. Cells are the starting point where both health and disease begin. When cells are treated well, they are able to function in health.

What action we take with this knowledge is our choice.

Chapter 6
Cells Know How to Function

Cells are intelligent, and in order to cooperate with that intelligence, we have the opportunity to learn what they need and to respond. An initial desire to cooperate in that way is important. After all, if we don't want health, the game is over. However, if we do desire health, first we need some information.

Cells become deficient when denied nutrients, and while willful cooperation engenders positive energies (emotional, mental, spiritual, physical) in the body, and the presence of these energies is supportive of cells, the missing information is the specific nutrients still needed. Let us begin with addressing additional needs during physical exercise, as this directly affects nutritional needs.

During exercise, the needs of cells change, and depending on the amount of exercise, their needs usually increase. Guidance and recommendations from a sports dietitian or other fitness professional, as well as reliable reading on the subject, is important for those exercising beyond moderation on a regular basis. It is essential to work appropriately with the body. In extreme sports, nutritional modifications must be calculated very particularly to match performance needs. Although a slight over supply of some nutrients may be favorable, balance is optimal.

The question of over supply is important. The most obvious potential for over supply regardless of exercise is calories, that is, simply, too much food for what we actually need. Americans actually require far less food than most of us think. Large servings of rich, creamy, fried, and sweet foods and beverages, both carbonated and non-carbonated, have become comfortable habits. However, while it has become a way of life, these choices may have a big price tag later. The Mediterranean Diet (see Appendix) has the best record of accomplishment for health, and it calls mostly for fresh foods and very few sweets.

Most of us know that calories in over supply store in the body as fat. Vitamin A and the mineral iron, which are easily found in foods, are also stored and are unlikely to reach toxic levels. However, with the use of dietary supplements, we could potentially reach toxic levels, so how we choose supplements is critical.

There are many nutritional supplements on the market, and the biggest difference between them is the amount of scientific study that has gone into their development. This type of exhaustive study means investigating not only the appropriate nutritional formula that demonstrates continuous support of human cell health, but also the necessary manufacturing process that delivers the same purity and potency every time.

Since supplements are manmade, the transfer of nutrients from the original plant form to the packaged form deserves examination. There is a lot to wonder about here beginning with the harvesting procedure of live plant parts. After harvesting, these parts are then stored, transported, and received, during which time standards for these procedures vary greatly between business organizations. Discerning manufacturers confirm that the received item is the correct plant and part according to a certified inspector, and then test for contamination. At this point, nutrients are extracted in a protected environment, tested for potency and purity at every step of production, mixed according to the formula, measured for the form and amount listed on the product label, tested for dissolution, and finally packaged so the product in the container matches exactly what is on the label. Supplying active ingredients that match cell needs requires high standards and close monitoring that challenges any manufacturing organization guaranteeing consistency in every container.

The intelligence of our microscopic cells is staggering, and immunologists know that human cells perform thousands of nutrient transactions in a split second. Humans can work to sustain the vital environment in which they thrive by providing the nutrients that they have shown contribute to their health.

Our cooperation in providing cells with what they need is done through a combination of food and supplements. As humans become more aware of what is involved and take action, cell health can occur reliably and deliberately.

Chapter 7
Allow Cells to Choose

Allowing is an act that initially calls on a willingness to trust something or someone including ourselves. Allowing can also be based on chance for those willing to take a risk before studying what is involved. For those who like to be informed, the word trust means "assured reliance on the character, ability, strength, or truth of someone or something." Trust then involves a relationship with one's self, or between two or more people or things, and that relationship can be neutral, positive, or negative. The point is that this connection calls for a decision to permit or not permit an action, and it is based on trust.

Given that cells are a creation of Nature, the decision to trust their activity involves Nature itself. Are we, as humans, in a position to override Nature? After all, we are creations of Nature, and cells are basic to our body forms. Can we trust our cells to do their work and, further, to know what they are doing?

Scientific studies show that we have hardly scratched the surface of understanding the genius of Nature. Can anyone really overtake something that they don't truly understand? Dominance may have appeared to work well in the past, however, in the long term, we have brought on a lot of manmade problems with this approach.

At this point, cooperation seems a reasonable and worthy pursuit. Naturally, scientific knowledge is critical for building a solid base of understanding toward a realistic alignment with Nature and, fortunately, we continue to advance our knowledge every day. In addition, as creations of Nature ourselves, we have an innate ability to harmonize with all of Nature. It remains our choice to execute this ability.

Cellular intelligence displays itself throughout the thousands of interactions that occur every split second within a single cell. During that second, a cell utilizes and transforms the nutrients it needs when they are available in a network of interdependence and inter-relating. An extremely abbreviated description of cells functioning together effectively for a single purpose is:

1. Cooperation with each other for the good of the whole body.
2. Constant communication in order to cooperate effectively.
3. Awareness of the vital and interdependent function of each cell to the whole.

All day long and through the night, this activity continues as consistently as our breathing and heartbeat. We have the opportunity to participate actively in this process by extending our own cooperation.

Cooperation

When the thought of cooperation comes up, many of us ask, "what's in it for me?" Although that is not a very cooperative question, it does lead to an important component of cooperation.

Webster's New Collegiate Dictionary defines *cooperate* as "to associate with another or others for mutual benefit." The word *mutual* in that definition is important, and sheds light on the fact that someone, or something, else is involved. The question about cooperating then becomes, "how is this activity benefiting all participants?" In this question, the cooperative benefits to others are vital, and what follows as the rewards with this type of relationship, besides functional results, are good feelings mentally, physically, emotionally, and spiritually.

While these rewards are well deserved, once they are achieved it is a mistake to assume that they will sustain themselves without continued attention. If this perspective develops and prevails, the cooperative energy drops, and the drop in energy will soon have us back at the beginning, prior to the rewards, and at a loss as to what caused the collapse. Results must be valued enough to keep feeding the cooperative source that created them. In a similar way, only with continual attention are genuine feelings of success sustained. For example, when we deliver respect, love, or kindness to ourselves, or others, the feelings of success that come from those offerings sustain only as we continue to deliver them. In doing so, it makes for a lasting shift in our habits and way of being.

Shared thoughts and goals that contribute to our productivity in working together facilitate sustained cooperation as well. Such thoughts with regard to cells and nutrition include:

1. Purchasing mostly fresh foods with reference to choices within the Mediterranean Diet.
2. Using nutritional supplements with a science-based formulation and manufacturing process.
3. Exercising the body and experiencing the good feelings that accompany cooperation with Nature.

Choosing cooperation, and learning to flow with this choice, can effectively enhance our innate capabilities and performance in life. When we look at cellular cooperation, as reported by scientists through microscopes, we see that human beings are inseparable from the vast matrix of continuous interactions; cells respond and cooperate with our choices and influences.

Finally, in response to the question of, "what's in it for me?" The answer is wellness and the rewards that go along with that both for ourselves and for everything, and everyone, around us; everything, meaning that the world environment is included.

Even when we provide an abundant supply of nutrients to the body each day, not all of them are needed, depending on our activities. An extended component of cooperation with the environment is that nutrients that are neither stored nor absorbed pass through the body and back out into the world, and this re-entry to the soil completes the circle of Nature. Once surplus nutrients pass from the body, they enter the public system where they are treated, hopefully appropriately, and returned either to the soil or to the water supply.

Chapter 8
Live and Love Life

As a desirable way to move through this world, the "live and love" lifestyle has very few adversaries. In fact, I have never met anyone consciously opposed to it. For a long time, I myself did not realize what it even was. Yet during my pursuit of health and joyful participation in life, this was the lifestyle with the most energy; and I found myself on a journey to the home of the heart and soul.

Over the years, with a curious and ever more delighted experience of what is available in life, I engaged in slowly removing the limiting beliefs I had held for so long; and these were thoughts that with closer examination could be relinquished as incomplete in the light of new knowledge and insights along my journey. Indeed, my travels brought me "face to face" with myself and my view of who I am, which by the way, was very unclear at first; and these creative experiences were ultimately priceless in value although I did not see it that way at the time. Situations like these were pivotal points at which I had a choice to go forward, look the other way, and take on more activities to keep my attention, or cover it up with disbelief and denial; and I chose to go forward by looking into what was happening.

These issues called upon my personal stability, internal clarity with myself, and having my "feet on the ground"; and it gave me the chance to examine my views of my self-acceptance, self-love, and self-respect. Finding the truth about these for myself and then cultivating ownership of them within my being was vital; and I noticed that this had to occur before I could genuinely see them in others or extend them to others. This shift in my thinking was extremely gratifying to me.

Awareness that relations with myself were fundamental to building relations with others, prompted me to experiment a little; and I wanted to gain more insight. For example, I wanted to see what is natural to happen in a group of friends when I genuinely experience and show love and acceptance for myself including ownership and belief in the truth of these views. What I noticed by doing so, was, naturally, that I liked myself better and experienced more confidence; and because I liked myself better, I was more accepting of others and had more fun. In fact, after a while, it appeared that consciously owning and expressing my own positive nature was actually a precedent for my extending the same outward to others.

Although this was a new discovery for me, it really is a principle that has been around for a very long time; and certainly, it makes perfect sense that we can only give away what we already have; and in this case, the added piece is that we can only see good things in others when we can see them in ourselves first. Taking this awareness a step further, I decided to add more practice; and it was an enlightening experience to notice that viewing others from a genuinely positive mindset, and genuine expression of it, typically prompted similar expression in return.

Being in control of this innate guidance system for my own life, making individual choices for my being in this way, remains, to this day, a gift for which I am immeasurably grateful; and it enhances my intention and willingness to extend similar awareness and opportunity to others.

Recognition of this personal choice made the process of discovering the truth behind other issues of discord in my life easier; situations previously seen as fearful or baffling could now, by choice, be viewed with an attitude of exploration. New understandings of previously unknown material could be revealed and received in a spirit of adventure, openness, and creativity. Although in the short run it may have been upsetting to learn the truth of the matter, in the long run, it was a very enlightening and an energizing way to live in the moment.

Aspects I explored then for myself were forgiveness, appreciation, personal freedom, and service among others; all of these surfaced over time for clarification so my health and liveliness could be moved forward; and by consciously relinquishing the old habits and attitudes, I was free to "choose again."

Clearly, these shifts in thinking took place over several years and my ownership of the choice to make them was significant; after all, I was essentially choosing to *be* the effect I desired myself, first, before I could see it happen externally.

Making these shifts, not surprisingly, required patient contemplation and meditation; and yet, the harmonious benefits and new perspectives could be integrated consciously, responsibly, and accountably; and I wanted to feel good about them. Typically, after such consideration, the new "ground to walk on" made sense to me; and it really did feel good in my body and mind; and I appreciated how these shifts in thinking also had a beneficial effect on my cells.

Such peaceful experiences, both in the body and the mind, could not have been more welcome energetic vibrations; and I noticed that a joyful state of mind was present when I chose viewpoints that aligned with what is natural in life.

All of these realizations and more eventually merged into a way of living that still continues to grow and expand each day; and this lifestyle is one of authenticity, that is, *being* as we are designed to be by Nature. The more I realized that being in harmony with the natural order of life *is* life, the more I loved being alive; and it seemed like this was a principle I had been hearing for years, "Be who you are created to be." And in allowing an experience of this principle, even for just an instant, I became aware of a vast infinite opportunity for actually truly being who I am created to be; and just that brief moment alone, allowed me to open to the expanse and spaciousness available for human beings; as I experienced myself in this way, I can now see others with the same appreciation as to who they are truly created to be by Nature as well.

Naturally, it makes logical sense, that being as we are born to be is Nature's design; and although this is how it is in life, amazingly, as part of this gift of being who we naturally are, we have individual choice about what we do with that information. In other words, it is our personal choice as to whether or not we implement this already natural opportunity to be the presence of who we are born to be. It is a choice that for me, is easy to make.

From moving through all those experiences, the concept of choice was always the bottom line; and then I learned it was also the top line and everything in between. I can say for sure, the best thing to take ownership of, next to love and life, is choice. Choice encompasses everything that we think about and do; it is our freedom to express ourselves and be who we are.

When we look at it honestly, we see that life turns out the way we choose it to be.

One of the choices presented throughout this book has been our choice in relationship with cells. These microscopic parts of nature are vibrant and vital partners if we make the choice to relate to them that way. Though cells are invisible, we have the option of becoming fully aware of them and treating them well, and our demonstrations in this regard show how much we love the *all* of life. Cells function at the most fundamental level, and life, in total, depends on their aliveness.

Such vibrancy brings forth creative energy, and our cooperation in response engenders the same vigor. Although cooperation can become procedural, it is most alive when the heart and soul are willingly involved. However, in some situations, it could conceivably include both. The important part is the energy. Creative energy engages the honesty and clarity that comes from the heart. In my experience, activities are more fulfilling when our participation comes from the heart. Passion and love for the work we do are engagements that are both satisfying and gratifying. It is the nature of our being to love participating fully and being alive.

Those who have dabbled in wellness for a long time know that it takes a way of thinking that permits openness to new perspectives. Indeed, it involves looking at old situations and habits from new points of view that ring true with the heart, and the resulting resonant feeling is free and natural. Occasionally, such resonance requires both guidance and giving ourselves permission to have perceptions that are more expansive.

One of my biggest awakenings in life was when I discovered that I actually had no idea what my fellow human beings truly needed—after food, clothing, and shelter which are observable needs—other than love. Having viewed most of my life from the opposite perspective, meaning I thought I *did* know, or could at least make a good guess, it was astonishing and yet relieving to find out the truth.

I found out, of course, how off the mark I was with my preconceived notions.

At that time, I had not examined my own thoughts, nor been exposed to a broader thinking. It was a period where my judgments of me, and others, were unloving. In fact, they were controlling, arrogant, and flat out uncooperative. While getting to the root cause of such perspectives was hard work, looking back I realize it was worthy of my time as I actually gained immensely. Unfortunately, my agony reached such a high level that I nearly drowned in it before I took action. Exterior presentation and drama could not save me any longer, so I finally allowed myself to see an option that I was not open to before. I looked at a bigger picture, and a more realistic view of life, to understand some of the causes of my distress.

A typical response to my difficulties during those times was to patch them up with whatever brought immediate relief. Naturally, with this type of response, the origin of the problems remained in place at the cellular level where disharmony takes its toll. I know now that a patch applied at the surface of a struggle leaves the underlying cause to fester and recur in the same, bigger, or different form. Harmony is what we experience only if we choose it, commit to it, and envision it with passion. The adventure that follows is alive with responsibility and more choice.

Examples of harmonizing energies from the heart are acceptance, love, and compassion, which can also be viewed as being in alignment and resonating with the natural order of life; and these energies stand up in every day life regardless of how we misinterpret, misunderstand, and even misrepresent and deny them; they remain unchangeably beneficial. It is this internal work of building harmonizing relations within ourselves that creates a space for similar harmony to be extended to the external world. All these are gratifyingly pleasant vibrations in a broad context, and the path to implementing them is a journey of the soul worth both launching and living. The question as to the steps for the launching the journey remains.

Although healthcare practitioners are highly skilled and trained to be of assistance, individual responsibility is vital concerning healing. Opinions, examinations, and interpretations of data vary greatly, and while experienced practitioners definitely contribute valuable information, guidance, and service, each individual is ultimately the biggest influence on their own health. The responsibility and choice for health is with each of us.

Choices regarding food habits are an obviously individual activity that influences our quality of life. These choices affect our health and our environment simultaneously and, indeed, are either cooperative with Nature or not. Thus far, we have collectively not done well with this choice. While there are small numbers of people that have educated themselves and are experiencing the benefits, the shift in viewpoint is personal and everyone is held accountable. Humans benefit from the good choices because the cells that support our aliveness benefit. As we expand the shift in our habits, the environment, including the air, water, plants, birds, fish, and all the land animals, benefit.

Readiness to participate in this way may mean contemplation and considering new and experimental viewpoints of life and relationships. In order to integrate and accept a new choice, the advantages often need to be experienced firsthand.

I first became familiar with what transparency meant to relationships while catching up with a circle of friends. Transparency means *allowing* intimacy. It is the "look into me and see me" aspect of a relationship; it is sharing the truth of the world and ourselves in it. Friends who share honestly, safely, and openly embrace the world as it is naturally, before choosing it to be different than it is, that means without sugar coating and without lying to protect ourselves, or misrepresenting ourselves for any other reason.

Since feelings of fear and love do not exist in our minds at the same time, the world shows up for us the way we freely choose it to appear consciously or unconsciously based on which mindset we allow. Our *conscious* awareness of making choices from a mindset of love, which is the mindset that aligns and supports Nature, also sustains life as it was created to be; at the same time, that love mindset allows us to make choices exclusive of fear. Indeed, a mindset based in love and the associated feelings of compassion, respect, and many others, allows manifestation of the joys of aliveness.

What my friends and I found with transparency was how similar we really are at heart and how interdependent we are, and it was enlightening. Sharing the stories of our unique experiences and journeys enriched us both as a group and individually. We discovered that the initial risk of what felt like vulnerability led to the experience of invulnerability and healing. By allowing the power and energy that rests in love, we experienced harmony.

Between both friends and cells, the attitude of cooperation is the same. It is one of mutual benefit and harmony. Cells are alive and vital and thrive on high energy just as we do. Every day, we have the opportunity to strengthen the awesome network that is inseparable from us. The relations we develop with ourselves, and others, are strong steps toward health that extend to everyone, and everything, in the universe.

Chapter 9
Play

Since life is Nature's game and we are inseparable from it, we might as well play the best game we can. If we flow with life, we are in harmony. When we do not, we get struggle and strife. It seems like an easy choice to make.

Initially for me, the idea of harmonious relations every day was inconceivable. Even harmony for just a portion of a single day was a new and creative adventure. I could only imagine the feelings that would come with continuous transparency and trust for mutual goals and benefits. Reaching goals together, vulnerability, and just plain telling the truth seemed like a game that did not exist. Games I had been immersed in previously were pretentious, dramatic, competitive, and contrived, promoting posturing and little genuine freedom of expression. I admit that when I got a glimpse of the possibility of harmony and freedom, especially coupled with great health, I didn't believe it. In fact, I was nearly one hundred percent skeptical. It took some careful study to see that it was real, and when I finally saw what life's game was all about, I chose to take responsibility for my part in it. The power and excitement of the game designed by Nature compelled my participation, and my first activities involved envisioning cooperation, collaboration, friendship, and good health.

New insights, relations, and opportunities became available and, since then, continue to serve me well.

Helpful lifestyles that work for mutual benefit every day are a celebration that flows with the natural order of the universe. We become free to express ourselves and build the lives for which we were meant once we have created harmony within ourselves.

It all begins with a vision to manifest a desire. Team coaches are aware that it is the players, themselves, who "start their own engines." Players must have the desire and commitment, and a coach supports them in reaching their goals. When we choose to envision goals in the present moment, as if we have already accomplished them, passion and grace work to open the space for insight and opportunity to follow. We are responsible for "starting our engines," and we are the ones playing the game so that our visions become reality. Everyone has the same choice and capability.

The journey that follows may contain proverbial stormy times and rough waters, but it is exciting and creative as well. Allowing goals to materialize is also a choice, and when we reach the destination, we owe ourselves an appreciation and celebration of the entire journey. Often the most trying times are the most valuable.

Playing life's game with skill is admirable, and it happens more easily when we love what we are doing. The quality of our work is higher because our hearts and passions are present. Team members working together, in harmony, toward goals have a synergy that benefits everyone, including each other. By extension, the presence of harmony affects the individual cells of the body, the body as a whole, other individuals, families, communities, and even the universe.

It makes sense that harmony has such effects because love is the native environment that is home to all. Love is the language of the universe. Joy and love are high vibrations that are integral not only to cooperation, but also to service. Although we frequently interpret service as receiving, giving service in alignment with Nature's design brings about perfect harmony, or flawless internal calm, a quiet stillness and peace in the present moment. We are free to choose such calmness at any time, and it is the perfect setting for the joyful expression of our natural spirit—play.

Starting with the food produced from the soil, cells absorb nutrients and transform the energy in the process. Aligning with the transformative energy and power of Nature brings harmony, and separation brings discord. Knowledge of the game is key to playing, and it is your choice to play well and celebrate the joy.

Human beings are, arguably, the most intelligent life form on the planet, and it is imperative that we accept this fact and welcome our option to master the game. The potential has always been within our nature.

Since health and disease are opposites, it is our state of mind, our openness, and fearlessness to encountering all of life that makes the difference and enables perfection. Thoughts from the heart and soul generate a state of mind that is supportive of life. Unfortunately, ego-based thoughts do not. In every minute, we are choosing the life we desire and getting the results. Awareness of the origin of our choices, whether as a connection with the wisdom of our heart and soul or egotistical aspirations, is in our hands.

For fifteen years, I have admired the magnificence of human cells and believe that they deserve our full cooperation and awakened state of mind. Each of us can play this phenomenal game of life well when it is our vision and choice to do so.

Part Two

Part Two is the journey to health and joy told in pictures.
Because pictures communicate visually and in stillness,
I included them so that readers may choose to put themselves
in each picture. Often the greatest value in reading is when
we can see our own lives and thoughts moved forward
toward new ideas and insights. Having selected the pictures,
I included possible reflections on them, however,
the reader's own imagined journey is the most valuable.
Enjoy!

The Journey of Desired Change with Reflections on Pictures

Manifesting change is an inside journey as it comes from within each of us. Changes that happen on the outside are because we have changed the way we see things from the inside. Try it for yourself and see. As soon as we choose to look at things from a different perspective, we discover they actually are different than we had originally thought.

It is my belief that pictures offer the reader an opportunity to step into the scene themselves and experience the journey alongside the pictured characters. Readers may find it valuable as an imaginative and constructive process for considering different perspectives on the subject of creating health and joy.

The reflections that go along with the pictures directly relate to the themes presented in the chapters of Part One. It is possible to reconsider long held beliefs in the light of new information and choose to relinquish them in favor of the new insights. I have discovered that there are many reactions related to this process. Some feel relief and joy, while others experience resistance, denial, shock, or disappointment, among other emotions. Although initial responses eventually pass, I have noticed that it is in their very occurrence that internal changes come about. This effect is due to a new awareness of previously unknown aspects of life, for example, "aha moments," or new realizations, may occur that are a departure from familiar beliefs; and resonating with these fresh points of view exposes a bigger picture and launches creative thinking.

In the following pages are images of people moving through the process of change toward manifesting a vision. These are examples of people demonstrating, through body language, their emotions, attitudes, and thoughts. Additionally, there is an illustration of a cell depicting the energetic attraction and silent demonstration of what I refer to as "cell talk."

Transforming our viewpoint and gaining understanding is a journey based on our trust in life. This means proceeding at our own pace, truly feeling the value of, and appreciating, the journey. Vision manifestation is an exciting and expressive dance with Nature's game and, taken patiently, step-by-step, it offers joy, awakening, creativity, and awareness of new and more powerful states of mind.

Reflections on Chapter One

A Vision of Health and its Celebration

In this picture, a man and woman envision good health and celebrate it with joy and gratitude. Within their vision, they imagine reveling with friends as they acknowledge the activities that have helped keep them on track and contributed to their goal. The image in the scene could be representative of anyone having a vision for their life such as teachers, athletes, politicians, scientists, artists, students, business or health professionals, volunteers, support personnel, community workers, or retirees. An important part of this picture is that they are standing in an area where it is quiet and pleasant, a setting where they are free from distraction and able to imagine whatever they want.

This particular setting shows a home in the distance, surrounded by trees. It is possible the air is clear and the water is pure. The land has streams and trees in harmony with Nature. Although the everyday circumstances of the people in this scene may be far different from the countryside they are presently in, there is no reason they cannot envision, with great passion and intention, the life they desire. They can then use that awareness to their advantage. Surrounded by the harmonious vibrations of humans living cooperatively with the land, they can appreciate the intelligence of Nature and be optimistic. They project such a strong picture of the life they desire that their minds and bodies vibrate with the feeling of being in that picture right now. The attraction is strong enough to compel a commitment on their part to see it through to the end. Their vision is realistic, including feelings about the relationships among players who have been tested, off and on, throughout their lives. Players who aligned with their own vision and goals supported them, and they are present to celebrate their individual joy, as well as the happiness of the group. Together, they have learned how to cooperate with others in the world, and they love celebrating in vibrant health and joy.

Reflections on Chapter Two

Investigating the Source of Health, and How to Influence It

In this picture, a woman holds the vision of having good, life-long health. She examines the project as represented by the sphere and seeks to understand exactly what health is and how she can influence it to her greatest benefit. Her intention is to gain information about the origins of health, more specifically, about the starting point for wellness or disease in the body.

The fact that cells are microscopic is interesting to her, although unfamiliar. She has never actually seen a live human cell under a microscope. What is important to her at this point is that the cellular level is where both health and disease begin. This fact intrigues her as she begins to recognize the importance of such tiny parts to the functioning of the whole. Before this, she had not been aware that each cell is an enormously intelligent, fully functioning unit on its own. She sees now that cells cooperating interdependently is a phenomenal thing. As the new information wakes up her mind and senses, she realizes that she has a major part to play in promoting the direction of health in her body. The function of the trillions of cells in her own body is interdependent with what she does. She now has the choice to cooperate and support their best performance, or to neglect them. With either choice, her cells respond accordingly and either thrive or try to survive the best they can.

This woman is torn between putting her attention on other matters in her life that demand her time, or spending more time learning about manifesting her new vision for health. She has many ongoing activities in her life that cannot be ignored as they are important and rewarding. She considers the possibility of putting health "on the back burner" because money, family, school, business, and social responsibilities already fill her day. She reasons that each day only has so many hours, and perhaps she should break her commitment to health.

However, she still has her vision, along with the desire to love being alive. She wants to be free and able to participate, alert and expressive, both in body and mind. Her passion for this lifestyle has not drifted, however these desires carry a lot of weight. After considerable thought, she decides to share her vision, and its value, with others and to stay on the track to learning more.

Reflections on Chapter Three

Reconciling the Environment as Critical to Health

A man sits on a horse, extending a sphere that represents the condition of the food supply. Because the situation is far bigger than he is, he comes with additional power and strength by riding the horse, as he makes the gesture of presenting information. What he has found in his journey is that maintaining good health involves far more than he realized.

Unfortunately, the sphere reflects a sad story of the food industry—not irreversible but nearly devastating. The evidence shows that humans have caused some major problems for themselves. However, his demeanor remains calm as he recites both facts and figures.

The importance of soil is the bottom line of his message and, prior to this, he had taken it for granted due to a lack of awareness. Now, he recognizes that the soil is part of a cooperative matrix in the universe that deserves respect for its service to all living organisms. Before, he had been ignorant of the facts but, finally, it makes sense, and he is relieved there is a way to help.

However, now that the truth has been revealed, the economics of it are distressing. The role each human being plays in the consumers and producers, money and health game is critical. He had not been aware of the importance of selecting certain foods because only recently have the facts caught his attention. Nutritional value had never been an issue. He knew that food was necessary for strength and the feeling of nourishment, but the actual nutritional value of food he saw as something other people talked about; it wasn't worth his time. Now seeing the difference that foods make to health, especially his own, he is doing his part to inform others.

Having studied the situation, he recognizes that times have changed in a dangerous way, and it concerns him because he knows the food supply is not offering what cells and bodies need. Even as it affects his own body, it saddens him further when he thinks of the bodies of children, which are exposed, almost entirely, to the convenience foods of today.

His message is one of conviction and commitment to being part of the change. As he travels, his momentum builds, knowing that the outcome of health affects everyone, and everything, on Earth, and that his message is urgent. He moves on with courage, aware that others make decisions based on what is within their current beliefs and experiences, and yet, what they choose affects everything on the outside.

He learns as he goes along and his vision for health drives him forward. He urges involvement, not only for his own well-being, but also for all the creatures of nature, including his own horse.

Reflections on Chapter Four

Discovering Who to Trust in the Industry

A man stands with his arms folded, presenting a self-contained attitude as he observes a sphere that represents the food industry. He is aware of how entwined the food industry is with economics and has decided to find out for himself the foods in supermarkets that are nutritionally valuable. He is not pleased with the problems we have created for ourselves, and he does not look forward to sorting through the overwhelming supply of processed foods to find the foods that are worth his money. He is optimistic, however, that there are foods that are not only good for every cell in his body, but also look and taste great.

He has a right to be angry about the chemicals in the soil that deplete nutrition, and he did not ask for all the processed foods. However, given the situation, he is willing to do what he can. As one person, it makes sense that he can not influence the whole industry, but he can influence his own health and tell others around him how they can influence their own health if that is what they choose.

The work it takes to sort and find valuable foods is new to him, but not difficult. He simply has to pay attention to the condition of the foods, ask questions, and read labels more often. The question of trust bothers him because finding a business that is trustworthy is really a search for an organization with integrity and management principles focused on priorities of health.

He is aware that trust, defined officially, is an "assured reliance on the character, ability, strength, or truth of someone or something," and he decides to trust himself for the time being. While the alert system used by the government to control labeling and food production problems is loose, at best, and yet acceptable given the enormous size of the industry, he sees that consumers would be better off if they can trust a producer's intentions.

Since even fresh foods do not have the nutritional value they used to have, he sees a gap that needs to be filled. Although he regrets that human errors have caused a separation from Nature's design, he moves ahead and educates himself about the dietary supplement industry so he can get the nutrition his cells need. Prior to this moment, he had not realized the differences between the numerous available supplements and, because they are manmade, trust in this arena is a big issue for him as well. He finally decides that science is the critical determinant, and that both the formulation, and the manufacturing process, must be scientifically based according to what supports cell health.

Even though sorting through information by asking questions is both necessary and beneficial, the man knows that occasional processed foods on his menu add adventures in flavor and savory dining. His highest priority is cooperation with cell needs because he likes to feel good and yet, occasionally, he chooses to balance the pleasures of processed food with his enjoyment of great health.

Since no one stands in judgment at the time of food purchase, each person is completely on their own to choose their products. This man recognizes that the market is where his decisions must be made because, once he gets home, his purchases will be enjoyed as snacks or meals.

This man has evaluated the situation, sees it is manageable and even fun, and takes responsibility for his food choices and his health.

Reflections on Chapter Five

Reversing the Damaged Food and Health System

The man in this picture is modifying his viewpoint in a major way. The sphere he is working on represents a new approach to protecting not only his own health, but also the health of others. With his increased understanding of the problems in the industry, he is taking action.

Now that nutritionally valuable foods have become familiar and pleasing to him, a bigger part of the problem reveals itself. Just as he has, many fellow human beings have overloaded their bodies to the point of harm with pollution from the air and excessive fat from rich, creamy, and deep fried foods. Unwittingly, they have damaged their cells as well.

His recent investigation of the system for supporting health was long overdue. He discovered that a very large percent of the population have contracted long-term illnesses due to various mistakes and practices, and he is angry. The situation both disappoints and frustrates him because it has gone on for so long and gotten so big. His desire to help others recognize the problem and take action has grown.

Although priorities were mishandled, and health overlooked, in the past, he is aware that he must be clear in his own mind about the source of the problem in order to help. He knows that the place to focus his attention and awareness is at the cellular level.

In this picture, he is hammering out a new approach to promoting health, which consists of rethinking his relationship to the system and transforming his way of being in the matter. He relinquishes old beliefs in light of new information, acknowledges and forgives mistakes, and puts reliable truths into place. This new approach aligns, happily and necessarily, with the designs of Nature.

Reflections on Chapter Six

Adding to Understanding by Looking Closer

This man examines a sphere that represents the function of human cells. He realizes that he needs to know more about these cells. In the past, their microscopic size was too small to hold his attention, and he felt that information on cells was a waste of time. He knew they were components of his body, it was just that other things seemed more important. At this point in his journey to health however, he sees that knowing more about cells makes a big difference in how he chooses to cooperate with them.

What reveals itself nearly overwhelms him. These invisible cells are sensitive to everything that happens, taking in the energetic vibrations of not only nutritional origin, but also of emotional, mental, spiritual, and physical origins. He recognizes that, in fact, his body is a virtual energy-transmitting device that uses trillions of cells in its process.

Prior to this awakening, there had been no reason for him to respect or admire cells. Now, he understands that not only nutrition is important for his health, but that his thoughts and attitudes also have a physical effect on his body. His regard for cooperation now takes on new importance since it manifests in his cells.

He releases previous viewpoints as he allows new information to transform and update his thinking. Now, he is interested in the opportunity to see illustrations of cellular activity. He knows scientists distinguish healthy cell activities through microscopic observation. As the intelligence of cells has become obvious to him, he wants to understand it better and, with new enthusiasm, he chooses to investigate further.

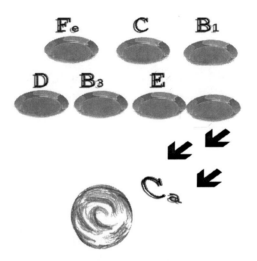

Reflections on Chapter Seven

Allowing for the Intelligence of Cells

This image invites the reader to allow a level of connection beyond the everyday senses of touch, taste, smell, hearing, and sight. Here, a cell in the foreground, represented in the sphere, shows the energetic exchange with vitamins and minerals. While this depicted exchange cannot be seen directly through a microscope, the results of it can be seen in the health of a cell when its needs have been satisfied.

The point of this picture is that there is a relationship between humans and our cells. It is not simply about humans cooperating with their cellular health, but also about an awareness of a cell's true service. Their connections with our energies in the world are silent, innocent, and allowing, serving us with complete acceptance and non-judgment. Conscious effort to join in oneness with their energy creates a quiet unity from deep within the soul. Ultimately, it is the aliveness that we crave.

Cells "talk" in their silent demonstrations. Although their part of the conversation is without sound, they make their communications clear through their behavior. Similar to witnessing a sunset or any other of nature's beautiful scenes, presence at such events invites stillness and opens the wisdom of the soul. In this state of mind, we can appreciate the miraculous transactions that occur during each second inside a cell and, with the help of a microscope, these results can actually be seen.

The cell in this image portrays a single action within the thousands it completes in a second. From the foreground, the cell has a virtual buffet of vitamins and minerals to choose from and, energetically, draws to it the calcium that supports its full function.

Indeed, cells need all the vitamins and minerals portrayed in this buffet, plus many others, in certain amounts. Additionally, they need elements of protein, fat, carbohydrate, and calories, again, in certain amounts. By design, cells know to choose exactly what they need from what is available in their environment; and it is known that three of these elements are stored by the body when there is excess. By design, Nature is always intelligent and perfect, and yet somehow humans overlook such loving genius thinking we have a better idea.

This picture depicts the ideal situation for a cell, as what it is looking for is, in fact, available. More often, though, nutrients are missing. When a cell needs a substance that is not available to complete a reaction, or when an element in the cellular environment, such as free radicals, is damaging them, cells automatically modify their functioning.

Reflections on Chapter Eight

Dancing in Alignment with Love and Life

While it would seem that the opportunity to experience life at its best has been fulfilled, there is always more. The man in this picture has noticed the positive effects of the changes in his daily living on his cells. Originally, he resisted creating such awareness, but now he is pleased that his vision of life-long health has led him to places he never imagined he would go.

With new admiration for life, this image portrays him as an open and active person expressing his joy. It is a demonstration of confidence and harmony with the activities and surroundings in which he lives. From the beginning, he had been committed to positive goals, and what he found out along the way was that by taking better care of his body, he had started feeling better. As he felt better, the momentum for his increased participation began to grow. He began feeling better about himself as a person, and that led him to a place where the difficulties and struggles of his life could be examined and resolved. With passionate feeling, he envisioned these struggles as already resolved, and began discovering truths that he had never imagined related to his issues. New ways of thinking and responding to the problems presented themselves and allowed him to experience greater clarity and self-acceptance. Liking himself better, his actions began to harmonize with those of others around him, and he experienced a calmness and peace that revealed the higher vibrations of life.

His journey has been an adventure of opportunities and new realizations and, although there have been many obstacles, there have been doors opened to him as well. In each situation, he looked closely to see the facts of the matter and, once he trusted his truth, he moved forward deliberately. Discovering and accepting Nature's design was the approach that allowed him to see what was truly there and, further, to align with it. Now, he dances joyfully while loving and living a vibrant and healthy life.

Appendix

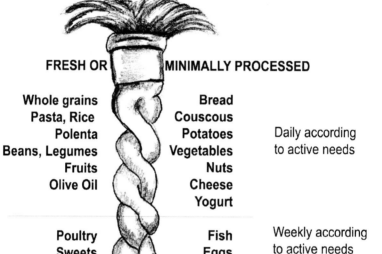

CPSIA information can be obtained at www.ICGtesting.com
Printed in the USA
LVOW041749070512

280696LV00004B/136/P